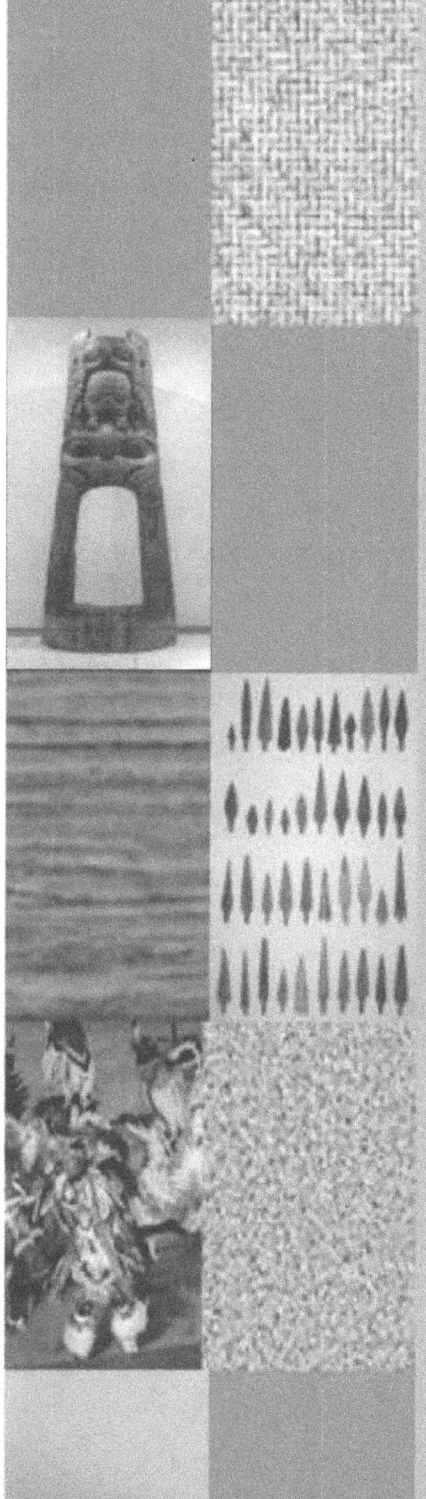

Indicators and Methods for Constructing a U.S. Tribal Well-being Index for Sustainable and Healthy Communities Research

Supplement 1 to EPA 600/R 12/023

January 2014

Photo credits for the cover

Baird, D. (2002, February 1). Indian 2 [American Indians dance during the Opening Ceremony of

the 2002 Olympic Winter Games on February 8, 2002, in Salt Lake City, Utah.]. Retrieved

from

http://commons.wikimedia.org/wiki/File:SLC_opening_ceremony_American_Indian_danc

er.jpg

Johnson, Adams, Hawk, & Miller (2011). Arrow heads [Cover picture from a USDA reoprt].

Retrieved from http://commons.wikimedia.org/wiki/File:Arrow_heads.jpg

Zienowicz, A. (2012). National Museum of the American Indian [National Museum of the

American Indian, Washington, D.C.]. Retrieved from

http://commons.wikimedia.org/wiki/File:National_Museum_of_the_American_Indian_-

Washington-_2012_(13).JPG

This document is intended to supplement U.S. Environmental Protection Agency's report EPA 600/R-12/023 entitled Indicators and Methods for Constructing a U.S. Human Well-being Index (HWBI) for Ecosystem Services Research (USEPA 2012). The domains described in the March 2012 report are also applicable to developing a tribal HWBI; however, specific metrics and data sources may differ from the original report and are identified in this document.

Acknowledgements

This supplement to U.S. Environmental Protection Agency's report EPA 600/R-12/023 was prepared by the U.S. Environmental Protection Agency (EPA) Office of Research and Development (ORD) National Health and Environmental Effects Research Laboratory (NHEERL) Gulf Ecology Division (GED) by the following 1.2.2.2 Task contributors:

Lisa M. Smith*[1]
Christina M. Wade[2]
Jason L. Case[2]
Linda C. Harwell[1]
Kendra Straub[2]
James K. Summers[1]

*Principle Investigator
[1] EPA/ORD/NHEERL/GED
[2] Student Services Contractor for EPA/ORD/NHEERL/GED

Abstract

A Human Well-being Index (HWBI) has been developed for the U.S. to help inform and empower decision makers to weigh and integrate human health, socio-economic, environmental and ecological factors equitably to foster sustainability. The integrity of the index structure is designed to be transferable to different U.S. population groups across space, time and demography. To demonstrate index transferability, American Indian Alaska Native (AIAN) specific data were evaluated for the calculation of a HWBI for AIAN and larger tribal populations. The applicability and integrity of the HWBI framework were maintained when using metrics scaled to assess well-being for AIAN and large tribal populations. Greater than 80% of the data available for a national AIAN assessment were specific to the target population, while the remaining data were derived from the general U.S. population. Full listings of domains, indicators and metrics for HWBI can be found in the U.S. Environmental Protection Agency's report EPA 600/R-12/023.

1.0 Overview

The Human Well-being Index (HWBI) is comprised of metrics that measure the well-being of the U.S. Metrics in the HWBI were selected because they were applicable to the majority of the U.S. population. Upon the completion of a population-based application for American Indian Alaskan Native (AIAN) populations, it was demonstrated that the integrity of the index can be maintained for different population groups residing in the U.S. (Smith et al. 2014). The majority of metrics applicable to the U.S. were also determined to be applicable to tribal populations. Potential modifications necessary to produce reasonably defensible well-being assessments were identified and HWBIs were calculated for the AIAN population and large tribal groups for the 2000-2010 time period. Greater than 80% of the data available for a national AIAN assessment were specific to the target population, while the remaining data were derived from the general U.S. population. Despite the utilization of non-target data, the AIAN well-being signature could still be differentiated from the U.S. HWBI, demonstrating the transferability of the HWBI approach (Smith et al. 2014). The degree to which the structure can be utilized for different population groups is dependent upon the quantity and quality of available data.

The composite HWBI is comprised of eight domains of well-being, described by 25 indicators and measured by 80 metrics (Summers et al. 2012; USEPA 2012; Smith et al. 2013). Indicator scores are calculated as the average of the corresponding standardized metric values. The domain scores are calculated as the average of the indicator scores and the final U.S. HWBI is an average of the domain values. The integrity of the index structure is designed to be transferable to different U.S. population groups across space and time. However, it is important to illustrate the strength of the index for estimating well-being for discreet populations because of the scale-independent nature of the HWBI approach. While the approach is transferable, there are potential modifications needed to assess well-being for the AIAN population of the U.S.

HWBI metrics for AIAN populations were reviewed and accepted based on three primary criteria: 1) metric was relevant to the AIAN population and data were available; 2) metric was relevant but no data were available; and 3) metric was not relevant to the AIAN population. Metrics were categorized based on results stemming from a review of the related data and where appropriate, suggested alternative metrics were identified.

Metric data were examined for ethnic specificity, completeness and analytical appropriateness. Only data that could be readily identified as AIAN-related were considered. Data records were encoded to differentiate between single ethnic and multi-ethnic identified information, AIAN and AIAN-mixed, respectively. Where tribal specific (TS) data were available, a Tribal Group identifier was included with the appropriate data. The TS metric values were aggregated into one of 38 Tribal Groups (Table 1). These groups correspond to tribal assignments for which the U.S. Census (2000) had available county population estimates.

Table 1 The 38 U.S. Census Tribal Groups.

Alaskan Athabascan	Crow	Pueblo
Aleut	Delaware	Puget Sound Salish
Apache	Eskimo	Seminole
Blackfeet	Iroquois	Shoshone
Cherokee	Kiowa	Sioux
Cheyenne	Lumbee	Tlingit-Haida
Chickasaw	Menominee	Tohono O'Odham
Chippewa	Navajo	United Houma Nation
Choctaw	Osage	Ute
Comanche	Ottawa	Yakama
Confederated Tribes of the Colville Reservation	Paiute	Yaqui
Cree	Pima	Yuman
Creek	Potawatomi	

2.0 AIAN Specific Data

Many factors were considered for metric categorization. For each metric, the collection method was identified as either random (e.g., exit polls) or complete (e.g., vital statistics). Metric categorization was based upon reported ethnicity, sample size and temporal scale data availability. All 80 metrics were categorized into one of six categories (Table 2).

Table 2 Description of each of the six categories used to classify HWBI metrics based on available data for AIAN and AIAN-mixed populations.

Metric Category	Category Description
I	AIAN population data suitable for annual analysis
TS	Tribal-specific population data available (Category I)
II	AIAN population data suitable for decadal analysis and AIAN-mixed population data suitable for annual analysis
III	AIAN-mixed population data suitable for annual analysis; AIAN population data unavailable or not suitable for analysis
IV	AIAN-mixed population data suitable for decadal analysis with more than 1 year of data available, annual AIAN-mixed population data not suitable for analysis; AIAN population data unavailable or not suitable for analysis
V	AIAN-mixed population data suitable for decadal analysis with only 1 year of data available. Additional years may be supplemented with alternative data sources or measures
VI	AIAN and AIAN-mixed population data unavailable or not suitable for analysis at any temporal scale

Approximately 65% of the HWBI metrics were classified as Category I, II and III with AIAN or AIAN-mixed population data available to assess measures at the annual scale (Table 3). Thirty-nine percent of the Category I metrics had tribal-specific identifiers. Decadal AIAN-mixed population data were available for 19% of the metrics in the HWBI framework (Category IV). The remaining thirteen metrics were classified as Category V with only one year of AIAN-mixed population data available for analysis or having insufficient data available for analysis (Category VI).

Table 3 Distribution of metric categories within the HWBI framework for AIAN assessments. Category I metrics shaded lighter gray indicate tribal-specific data availability; * indicates an alternative metric is suggested.

Domain	Indicator	Metrics									
Connection to Nature	Biophilia	VI	VI								
Cultural Fulfillment	Activity Participation	IV	V*								
Education	Basic Educational Knowledge and Skills of Youth	I	I	I							
	Participation and Attainment	I	I	I	V						
	Social, Emotional and Developmental Aspects	I	I	I	IV						
Health	Healthcare	I	VI								
	Life Expectancy and Mortality	III	III	III	III	III	III	III			
	Lifestyle and Behavior	I	I	I	III						
	Personal Well-being	I	I	IV							
	Physical and Mental Health Conditions	I	I	I	I	I	I	I	I	I	I
Leisure Time	Activity Participation	I	IV								
	Time Spent	II									
	Working Age Adults	I	II	IV							
Living Standards	Basic Necessities	I	I								
	Income	I	I	IV							
	Wealth	I	I								
	Work	IV	IV								
Safety and Security	Actual Safety	I	I	III	VI						
	Percieved Safety	V									
	Risk	VI									
Social Cohesion	Attitude Toward Others and the Community	II	IV	IV	VI	VI*					
	Democratic Engagement	I	II	IV	IV	IV	VI				
	Family Bonding	I	IV	IV							
	Social Engagement	I	I	VI							
	Social Support	I	V								

2.1 Category I Metrics

Category I metrics have a sufficient sample size (n ≥ 100) of AIAN data to examine the measure on an annual scale. Six domains and seventeen indicators in the HWBI framework have metrics classified as Category I for a total of 39 metrics (Table 4)(BLS 2012; Hazards and Vulnerability Research Institute 2012; NCES 2012; CDC 2013a; U.S. Census Bureau 2013).

The Center for Disease Control (CDC) was the data source for almost half of the metrics in this category, followed by the U.S. Census Bureau from which data were identified for 11 of the metrics. All of the tribal specific metrics identified were categorized as Category I measures.

The U.S. Census Bureau's American Community Survey allows public dissemination of records with tribal affiliation identifiers. National Center for Education Statistic's National Assessment of Education Process and the CDC's Youth Risk Behavior Surveillance System allow dissemination through restricted records available by special request. These sources provide data for fifteen of 38 Category I metrics, although the workable number of metrics may be less depending on available sample sizes.

Table 4 Metrics classified as Category I based on the suitability of available annual AIAN population data.

Domain	Indicator	Metric	Tribal Specific Metric Available
Education	Basic Educational Knowledge and Skills	Mathematic Skills	Yes
		Reading Skills	Yes
		Science Skills	Yes
	Participation and Attainment	High School Completion	Yes
		Participation	Yes
		Post-Secondary Attainment	Yes
		Bullying	Yes
		Physical Health	No
		Social Relationships and Emotional Well-being	No
Health	Healthcare	Population with a Regular Family Doctor	No
	Personal Well-being	Perceived Health	No
		Life Satisfaction	No
	Physical and Mental Health Condition	Adult Asthma Prevalence	No
		Cancer Prevalence	No
		Childhood Asthma Prevalence	No
		Coronary Heart Disease Prevalence	No
		Depression Prevalence	No
		Diabetes Prevalence	No
		Heart Attack Prevalence	No
		Obesity Prevalence	No
		Stroke Prevalence	No
	Lifestyle and Behavior	Alcohol Consumption	No
		Healthy Behaviors Index	No
		Teen Smoking Rate	Yes

Leisure Time	Activity Participation	Physical Activity	No
	Working Age Adults	Adults Working Long Hours	Yes
	Basic Necessities	Food Security	No
		Housing Affordability	No
	Income	Incidence of Low Income	Yes
		Median Household Income	Yes
	Wealth	Median Home Value	Yes
		Mortgage Debt	Yes
Safety and Security	Actual Safety	Property Crime	No
		Violent Crime	No
Social Cohesion	Democratic Engagement	Voter Turnout	No
	Family Bonding	Exceeded Screen Time Guidelines	Yes
	Social Engagement	Volunteering	No
		Participation in Organized, Extracurricular Activities	No
		Emotional Support	No

2.2 Category II Metrics

Category II metrics can be used to assess aspects of well-being for AIAN-mixed populations at the annual scale or for AIAN only populations at the decadal level (BLS 2012; CDC 2013a; U.S. Census Bureau 2013). Two domains, four indicators and four metrics were identified for Category II (Table 5). Category II metric data were found from three data source; the Bureau of Labor Statistics (BLS), the U.S. Census Bureau and the CDC. All measures were collected using random surveys.

Table 5 Metrics classified as Category II based on the suitability of available annual AIAN-mixed and decadal AIAN population data.

Domain	Indicator	Metric
Leisure Time	Time Spent	Leisure Activities
	Working Age Adults	Adults Who Provide Care to Seniors
Social Cohesion	Attitude Toward Others and the Community	Discrimination
	Democratic Engagement	Registered Voters

2.3 Category III Metrics

Metrics in Category III represent annual measures of AIAN-mixed populations only (CDC 2013b, 2013c) as the AIAN-only data were not available for assessments. Health and Safety and Security were the only two domains with Category III metrics (Table 6). All seven Life Expectancy and Mortality indicator metrics fell into this category as did one of the four metrics of the Lifestyle and Behavior indicator of Health. One of the four metrics of the Actual Safety indicator was classified as a Category III measure. The data source for Category III metrics was the CDC. All metric data in this category were collected using complete collection types.

Table 6 Metrics classified as Category III based on the suitability of available annual AIAN-mixed population data.

Domain	Indicator	Metric
Health	Life Expectancy and Mortality	Life Expectancy
		Asthma Mortality
		Cancer Mortality
		Diabetes Mortality
		Heart Disease Mortality
		Infant Mortality
		Suicide Mortality
	Life and Behavior	Teen Pregnancy
Safety and Security	Actual Safety	Accidental Morbidity and Mortality

2.4 Category IV Metrics

Category IV metrics allow for HWBI assessment of AIN-mixed populations on a decadal scale as they do not have sufficient sample size for resolution at annual or AIAN-only scales. Fifteen metrics were identified as Category IV metrics for a total of 15 metrics for 10 indicators across six domains (Table 7).

Table 7 Metrics classified as Category IV based on the suitability of available decadal AIAN-mixed population data.

Domain	Indicator	Metric
Cultural Fulfillment	Activity participation	Rate of Congregational Adherence
Education	Social, Emotional and Developmental Aspects	Contextual Factors
Health	Personal Well-being	Happiness
Leisure Time	Activity Participation	Average Nights on Vacation
	Working Age Adults	Adults Working Standard Hours

Domain	Indicator	Metric
Living Standards	Income	Persistence of Low Income
	Work	Job Quality
		Job Satisfaction
Social Cohesion	Attitude Towards Others and the Community	Helping Others
		Trust
	Democratic Engagement	Interest in Politics
		Trust in Government
		Voice in Government Decisions
	Family Bonding	Frequency of Meals at Home
		Parent-child Reading Activities

2.5 Category V Metrics

Category V metrics have data available for only one year for AIAN-mixed groupings. The four metrics classified as Category V are from the Cultural Fulfillment, Education, Safety and Security and Social Cohesion domains (Table 8). Alternative measures (Section 3.2) may be available to fill data gaps for Category V metrics.

Table 8 Metrics classified as Category V based on the suitability of one year of AIAN-mixed population data.

Domain	Indicator	Metric
Cultural Fulfillment	Activity Participation	Performing Arts Attendance
Education	Participation and Attainment	Adult Literacy
Safety and Security	Perceived Safety	Community Safety
Social Cohesion	Social Support	Close Friends and Family

2.6 Category VI Metrics

Category VI metrics (Table 9) have no data available for assessment. When Category VI metrics are identified within an indicator that includes metrics from other categories (I-IV), the problematic Category VI metrics have little influence on index integrity. Instead of excluding the problematic indicators, data for the general U.S. population serve as placeholder until data specific to Native American populations become available.

Table 9 Category VI for which no data were available for assessment.

Domain	Indicator	Metric
Connection to Nature	Biophilia	Connection to Life
		Spiritual Fulfillment
Health	Healthcare	Satisfaction with Healthcare
Safety and Security	Actual Safety	Loss from Natural Hazards
	Risk	Social Vulnerability Index
Social Cohesion	Attitude Toward Others and the Community	Belonging to Community
	Attitude Toward Others and the Community	City satisfaction
	Democratic Engagement	Satisfaction with Democracy
	Social E engagement	Participation in Group Activities

3.0 Data Selection and Index Calculation

Raw data from various publically available surveys were organized hierarchically by population group and temporal resolution (e.g., AIAN and Tribal grouping by year and decade); information on each metric can be found in the Appendix entitled "Summary of Metric Data Used in index Calculation." National AIAN and Tribal Group datasets were created by populating metric values from the most robust data (Category I) available according to the metric categorization process (Figure 1) and from existing U.S. HWBI metric data.

A single imputation method using the carry-forward technique (Zhang et al. 2008) was used to fill data gaps caused by temporal disparities found across data sources. Values were calculated based on existing data for the nearest year within a single population group. The AIAN data were scored using the U.S. HWBI procedure (USEPA 2012), with minimum and maximum values being carried over from the HWBI dataset to allow for comparisons between HWBI and AIAN scores.

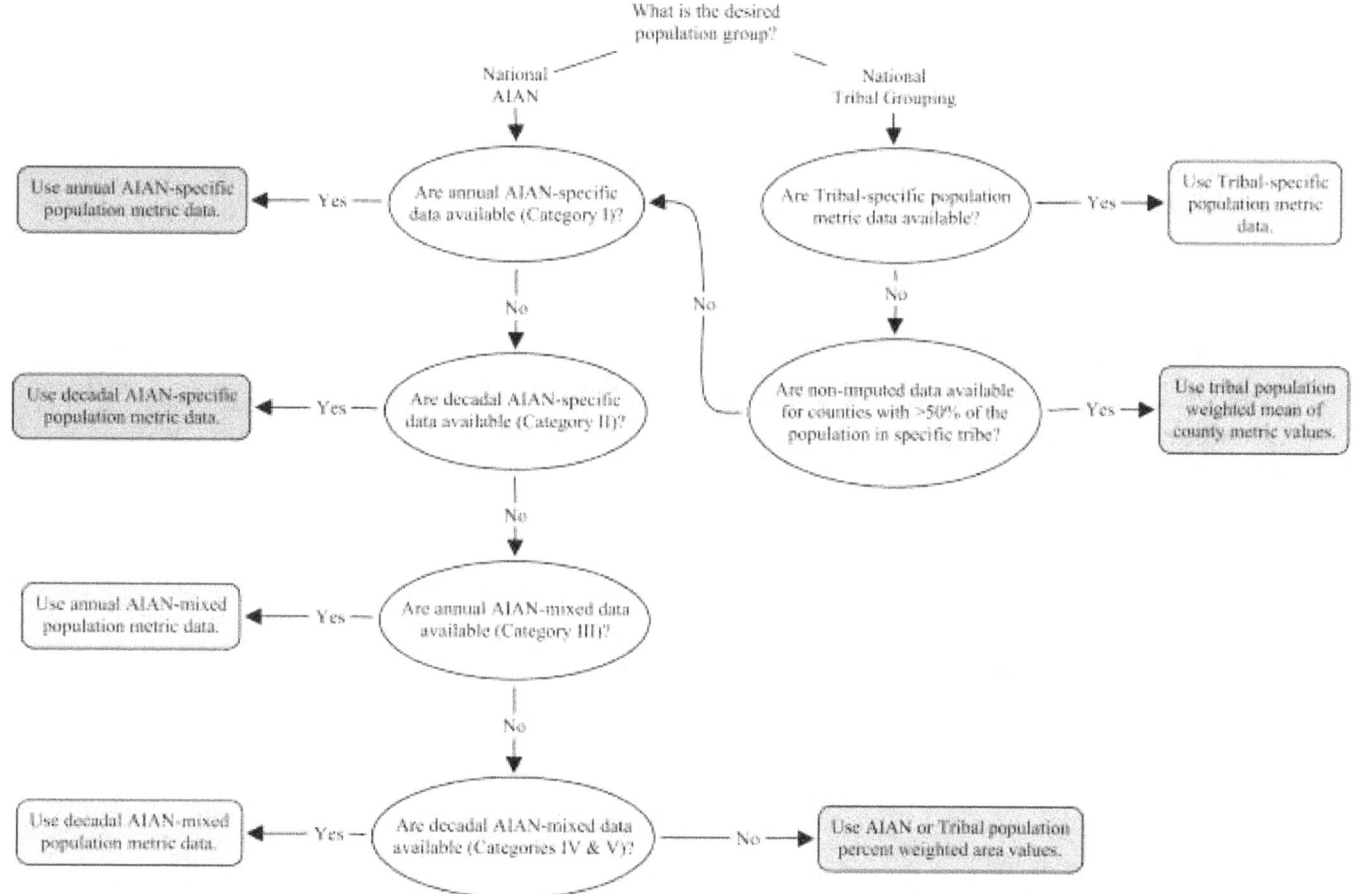

Figure 1 Process for selecting the most robust AIAN and Tribal Group data available for HWBI assessments.

3.1 Sources and Metrics

AIAN/Tribal metrics with data sources different from general U.S. population HWBI metrics

Domain: Cultural Fulfillment
Indicator: Activity Participation
Metric: Congregation Adherence
Category for Native Americans: Category IV
Metric Variable: TOTRATE
Source: American National Election Studies (ANES)
Source Question or Measurement: Frequency of Church attendance
Alternate Source: N/A
Years Available: 2000, 2002, 2004, 2008
Smallest Geospatial Level Available: National
Calculation Methods: N/A

Domain: Education
Indicator: Participation and Attainment
Metric: Participation
Category for Native Americas: Category I with Tribal Specific Metrics Available
Metric Variable: PARTNEDU
Source: Census-American Community Survey (ACS)
Source Question or Measurement: Attendance in a higher learning institution
Alternate Source: N/A
Years Available: 2000-2010
Smallest Geospatial Level Available: Country
Calculation Methods: Calculated as the percentage of people aged 18-24 enrolled in post-secondary education

Domain: Health
Indicator: Physical and Mental Health Conditions
Metric: Obesity Prevalence
Category for Native Americas: Category I
Metric Variable: OBESITY
Source: Center for Disease Control (CDC)-Behavioral Risk Factor Surveillance System (BRFSS)
Source Question or Measurement: NDSS variable ADJPERCENT, age-adjusted percentage of population aged 18 and older classified as obese (BMI≥30)
Alternate Source: N/A
Years Available: 2000-2010
Smallest Geospatial Level Available: County
Calculation Methods: N/A

Domain: Leisure Time
Indicator: Working Age Adults
Metric: Adults Working Long Hours
Category for Native Americas: Category I with Tribal Specific Metrics
Metric Variable: LONGWRKHRS
Source: Census-American Community Survey (ACS)
Source Question or Measurement: Number hours usually worked at all jobs
Alternate Course: N/A
Years Available: 2002-2009
Smallest Geospatial Level Available: County
Calculation Methods: Calculated as the percentage of employed respondents reporting that they work 50 hours or more per week

Domain: Living Standards
Indicator: Income
Metric: Median Household Income

Category for Native Americas: Category 1 with Tribal Specific Metrics
Metric Variable: HOMEVAL
Source: Census-American Community Survey (ACS)
Source Question or Measurement: ACS variable B25077, Median value of owner-occupied housing units
Alternate Source: N/A
Years Available: 2004-2009
Smallest Geospatial Level Available: County
Calculation Methods: N/A

Domain: Living Standards
Indicator: Income
Metric: Incidence of Low Income
Category for Native Americas: Category 1 with Tribal Specific Metrics
Metric Variable: POVERTY
Source: Census-American Community Survey (ACS)
Source Question or Measurement: All ages in poverty; Percent
Alternate Source: N/A
Years Available: 2000-2009
Smallest Geospatial Level Available: County
Calculation Methods: N/A

Domain: Safety and Security
Indicator: Actual Safety
Metric: Property Crime
Category for Native Americas: Category I
Metric Variable: PROPCRIME
Source: Bureau of Justice Statistics (BJS) – National Crime Victimization Survey (NCVS)
Source Question or Measurement: NACJD variables BURGLRY, LARCENY, MVTHEFT, ARSON, Number of burglary, larceny, motor vehicle theft, and arson offenses
Alternate Source: N/A
Years Available: 2000-2005, 2008
Smallest Geospatial Level Available: County
Calculation Methods: Calculated as the total (sum) number of property crimes per 100,000 people. Population estimates were provided by the NACJD (variable CPOPCRIM) and reflect the total population served by reporting agencies.

Domain: Safety and Security
Indicator: Actual Safety
Metric: Violent Crime
Category for Native Americas: Category I
Metric Variable: VIOLCRIME
Source: Bureau of Justice Statistics (BJS)– National Crime Victimization Survey (NCVS)
Source Question or Measurement: NACJD variables MURDER, RAPE, ROBBERY, AGASSLT,

Number of murder, rape, robbery, and aggravated assault offenses
Alternate Source: N/A
Years Available: 2000-2005, 2008
Smallest Geospatial Level Available: County
Calculation Methods: Calculated as the total (sum) number of violent crimes per 100,000 people. Population estimates were provided by the NACJD (variable CPOPCRIM) and reflect the total population served by reporting agencies.

Domain: Safety and Security
Indicator: Actual Safety
Metric: Community Safety
Category for Native Americas: Category V
Metric Variable: PRCVDSAFE
Source: Survey of Income and Program Participation (SIPP)
Source Question or Measurement: Gallup variable WP113, "Do you feel safe walking alone at night in the city or area where you live?"
Alternate Source: N/A
Years Available: 2009
Smallest Geospatial Level Available: County
Calculation Methods: Calculated as the percentage of people who responded "Yes"

3.2 Alternative Metrics

In order to maintain index integrity and capture the most holistic and comprehensive picture of a population, it is sometimes necessary to identify alternative metrics. The metrics utilized in the U.S. HWBI range in nature from individuals' perceptions (survey questions) to rates of occurrences of certain behaviors and outcomes in a population. When choosing alternative metrics it is imperative that both the qualitative nature of the information as well as the type of information is as closely matched as possible. We identified two domains for which alternative metrics are suggested: Social Cohesion and Cultural Fulfillment.

For the Social Cohesion domain, alternative data for the *City Satisfaction* metric in the Attitude Toward Others and Community indicator can be found from responses to the Survey of Income and Program Participation (SIPP) survey question "Overall, how satisfied are you with conditions in your neighborhood?" (U.S. Census Bureau 2001). The metric *Satisfaction with Democracy* of the Democratic Engagement indicator could be supplemented with measures found in American National Election Study (ANES 2013).

Domain: Social Cohesion
Indicator: Attitude Toward Others and Community
Metric: Neighborhood Conditions
Replaced: City Satisfaction
Metric Variable: EABSAT

Source: Survey of Income and Program Participation (SIPP)
Source Question or Measurement: "Overall, how satisfied are you with conditions in your neighborhood?"
Alternate Source: N/A
Years Available: 2001
Smallest Geospatial Level Available: National
Calculation Methods: Percentage of AIAN participants reporting very or somewhat satisfied with their neighborhood.

The *Performing Arts Attendance* measure of the Activity Participation indicator that describes Cultural Fulfillment is a Category V metric with a very limited amount of data available to address AIAN populations; therefore, necessitating an alternative. *Ceremonial Attendance* (NCES 2009) is the proposed alternative metric because it provides similar information and captures participation in a cultural activity specific to Native Americans.

Domain: Cultural Fulfillment
Indicator: Activity Participation
Metric: Ceremonial Attendance
Replaced: Performing Arts Attendance
Metric Variable: PERARTS (ALTERNATE)
Source: National Assessment of Educational Progress (NAEP), National Indian Educational Study (NIES 2011)
Source Question or Measurement: "Overall, how satisfied are you with conditions in your neighborhood?"
Alternate Source: N/A
Years Available: 2007, 2009
Smallest Geospatial Level Available: National
Calculation Methods: Percentage of 4th and 8th grade AIAN students reporting attending a ceremony at least once per year

The *Performing Arts Attendance* metric was replaced with the suggested measure, *Ceremonial Attendance* and a new Cultural Fulfillment domain score for the AIAN population was calculated. This resulted in a dramatic increase in the Cultural Fulfillment domain and Activity Participation indicator scores (Figure 2).

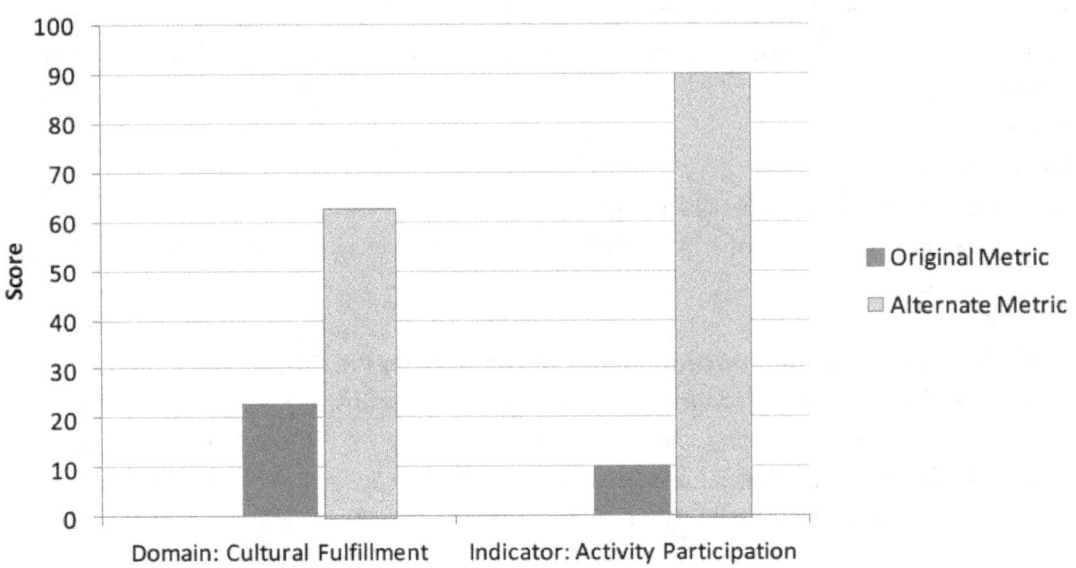

Figure 2 Comparison of the results of using an alternative metric for the Activity Participation indicator in the Cultural Fulfillment domain.

4.0 Concluding Remarks

The HWBI approach can be used to estimate well-being for Native Americans collectively with a reasonable level of confidence; however, the index may be less sensitive at tribal-specific scales as a result of non-specific data substitutions based on the general lack of publically available data for some key areas, limited spatial and temporal resolution of available data and inconsistent ethnic-specific identifiers in the data. To overcome limitations, data substitution using the described approach is the most robust method for scaling the index for AIAN populations. The degree to which the HWBI structure can be utilized is dependent upon the quantity of quality available data. However, as demonstrated, the framework can be used for assessing the well-being of discrete populations with minimal modifications at the metric level.

5.0 References

The American National Election Studies (2013). *Data Center*. Retrieved August 27, 2013, from

http://electionstudies.org/studypages/download/datacenter_all_NoData.php

Bureau of Labor Statistics (BLS). (2012). *American Time Use Survey*. Retrieved from U.S

Department of Labor website: http://www.bls.gov/tus/.

Center for Disease Control and Prevention (CDC). (2013a). *Behavioral Risk Factor Surveillance*

System. Retrieved from U.S. Department of Health and Human Services website:

http://www.cdc.gov/brfss/data_documentation/index.htm

Center of Disease Control and Prevention (CDC). (2013b). *Compressed Mortality File on CDC*

WONDER. Retrieved from U.S. Department of Health and Human Services website:

http://wonder.cdc.gov/mortsql.html

Center for Disease Control and Prevention (CDC). (2013c). *National Vital Statistics System*.

Retrieved from U.S. Department of Health and Human Services website:

http://www.cdc.gov/nchs/nvss.htm

Hazards and Vulnerability Research Institute (2012). *Social Vulnerability Index for the United*

States - 2006-10. Retrieved from University of South Carolina website:

webra.cas.sc.edu/hvri/products/sovi.aspx

National Center for Education Statistics (NCES). (2009). National. *Assessment of Educational*

Progress (NAEP), 2009 National Indian Education Study (NIES). Retrieved from U.S.

Department of Education website: http://nces.ed.gov/nationsreportcard/about/

National Center for Education Statistics (NCES). (2011). *National Assessment of Adult Literacy*

(NAAL). Retrieved from U.S. Department of Education: Institute of Education Statistics

website: http://nces.ed.gov/naal/

National Center for Education Statistics (NCES). (2012). *National Assessment of Educational*

Progress. Retrieved from U.S. Department of Education website:

http://nces.ed.gov/nationsreportcard/about/

Smith, L. M., Case, J. L., Smith, H. M., Harwell, L. C., & Summers, J. K. (2013). Relating ecosystem

services to domains of human well-being: Foundation for a US index. *Ecological*

Indicators, 28, 79-90.

Smith, L. M., Wade, C. M., Case, J. L., Harwell, L. C., Straub, K. R., & J. K. Summers. (2014).

Evaluating the transferability of a U.S. human well-being Index (HWBI) framework to

Native Americans populations. Submitted to *Social Indicators Research.*

Summers, J. K., Smith, L. M., Case, J. L., & Linthurst, R. A. (2012). A review of the elements of

human well-being with an emphasis on the contribution of ecosystem

services. *Ambio, 41*(4), 327-340.

U.S. Census (2000). Census 2000 Data for 249 Population Groups, including 39 Tribal Groups.

Retrieved from:

http://www.census.gov/aian/census_2000/census_2000_data_for_249_population_gro

ups_including_39_tribal_groups.html

U.S. Census Bureau (2001). *Survey of Income and Program Participation (SIPP)* 2001 Topical

Module. Accessed August 2013.

U.S. Census Bureau. (2012). *Current Population Survey (CPS): A Joint Effort Between the Bureau*

of Labor Statistics and the Census Bureau. Retrieved from U.S. Department of Commerce

website: http://www.census.gov/cps/

U.S. Census Bureau. (2013). *American Community Survey*. Retrieved from U.S. Department of

Commerce website: http://www.census.gov/acs/www/

U.S. Environmental Agency (USEPA). (2012). Indicators and methods for constructing a U.S.

human well-being index (HWBI) for Ecosystem Services Research. Report # EPA/600/R-

12/023

Zhang, S, J. Liao and Z. Zhu. (2008). *A SAS® Macro for Single Imputation.* Presented at Annual

Pharmaceutical Industry SAS Users Group. Atlanta, GA, June 1-4, 2008

APPENDIX

Summary of Metric Data Used in Index Calculation

DOMAIN	INDICATOR	METRIC	METRIC VARIABLE	LOWEST AVAILABLE SCALE		
				COUNTY	STATE	REGION
Social Cohesion	Social Support	Close Friends and Family	CLSFRNDFAM			x
	Social Engagement	Participation in Group Activities	GRPACTV			x
		Volunteering	VOLNTR		x	
		Children Participating in Organized, Extracurricular Activities	CHLDACTV		x	
	Attitude towards Others and the Community	Trust	CANTRUST			x
		Discrimination	DISCRM	x		
		Helping Others	HELPFUL			x
		Neighborhood Conditions	EABSAT	x		
		Belonging to Community	CLSETOWN			x
	Family Bonding	Parent-Child Reading Activities	CHLDREAD		x	
		Exceeded Screen Time Guidelines	WATCHTV		x	
		Frequency of Meals at Home	MEALS		x	
	Democratic Engagement	Voter Turnout	VOTRTOUT		x	
		Interest in Politics	POLINTRST			x
		Volunteering in Politics	POLVOLNTR		x	
		Voice in Government Decisions	VOICENGOV			x
		Satisfaction with Democracy	SATDEM			x
		Trust in Government	TRUSTGOV			x
		Registered Voters	REGVOTRS		x	
Education	Childhood Education and Care	Preprimary Education and Care	CHLDCARE		x	
	Social, Emotional and Developmental Aspects	Contextual Factors	CONFACT		x	
		Child Physical Health	CHLDHLTH		x	
		Social Relationships and Emotional Well-being	CHLDSOCIAL		x	
		Bullying	BULLY		x	
	Basic Knowledge and Skills of the Youth	Mathematics Skills	MATHTEST		x	
		Science Skills	SCITEST		x	
		Reading Skills	READTEST		x	
	Participation and Attainment	Participation	PARTNEDU		x	
		High School Completion	HSGRAD	x		
		Post-Secondary Attainment	UNIVGRAD	x		
		Adult Literacy	ADULTLIT		x	

DOMAIN	INDICATOR	METRIC	METRIC VARIABLE	LOWEST AVAILABLE SCALE		
				COUNTY	STATE	REGION
Connection to Nature	Biophilia	Spiritual Fulfillment	BEAUSPRT			x
		Connection to Life	ALLOFLFE			x
Health	Personal Well-being	Perceived Health	PRCVDHLTH	x		
		Life Satisfaction	LIFESATIS	x		
		Happiness	HAPPY	x		
	Life Expectancy and Mortality	Life Expectancy	LIFEXPCT	x		
		Cancer Mortality	CANCMORT	x		
		Infant Mortality	INFMORT	x		
		Suicide Mortality	SUICDMORT	x		
		Diabetes Mortality	DIABMORT	x		
		Heart Disease Mortality	HRTDISMORT	x		
		Asthma Mortality	ASTHMORT	x		
	Physical and Mental Health Conditions	Diabetes Prevalence	DIABETES	x		
		Cancer Prevalence	CANCER	x		
		Depression Prevalence	DEPRESSION	x		
		Coronary Heart Disease Prevalence	HRTDISEASE	x		
		Stroke Prevalence	STROKE	x		
		Heart Attack Prevalence	HRTATTACK	x		
		Adult Asthma Prevalence	ADLTASTHMA	x		
		Childhood Asthma Prevalence	CHLDASTHMA	x		
		Obesity Prevalence	OBESITY	x		
	Lifestyle and Behavior	Teen Smoking Rate	TEENSMK		x	
		Healthy Behaviors Index	HBI	x		
		Teen Pregnancy	TEENPREG	x		
		Alcohol Consumption	ALCOHOL	x		
	Healthcare	Population with a Regular Family Doctor	FAMDOC	x		
		Satisfaction with Healthcare	SATISHLTHC	x		
Living Standards	Wealth	Median Home Value	HOMEVAL	x		
		Mortgage Debt	MTGDEBT	x		
		State and Local Government Revenues	LOCGOVREV		x	
		Outstanding Public Debt	PUBDEBT		x	
	Income	Median Household Income	MEDINCOME	x		
		Incidence of Low Income	POVERTY	x		
		Persistence of Low Income	POVPERSIST			x

DOMAIN	INDICATOR	METRIC	METRIC VARIABLE	LOWEST AVAILABLE SCALE		
				COUNTY	STATE	REGION
	Work	Job Quality	JOBLOSE			x
		Job Satisfaction	JOBSATIS	x		
	Basic Necessities	Housing Affordability	HOMEAFFORD	x		
		Food Security	FOODSECURE		x	
Leisure Time	Time Spent	Leisure Activities	LEISURE		x	
	Activity Participation	Physical Activity	PHYSACTIV	x		
		Average Nights on Vacation	VACATION		x	
	Working Age Adults	Adults Working Standard Hours	NORMWRKHRS		x	
		Adults Working Long Hours	LONGWRKHRS		x	
		Adults who Provide Care to Seniors	SENIORCARE		x	
	Retired Seniors	Active Leisure	ACTIVESENIOR		x	
		Formal Volunteering	VOLSENIOR		x	
Safety and Security	Actual Safety	Property Crime	PROPCRIME	x		
		Violent Crime	VIOLCRIME	x		
		Loss from Natural Hazards	NATHAZLOSS		x	
	Perceived Safety	Community Safety	PRCVDSAFE	x		
Cultural Fulfillment	Activity Participation	Congregational Adherence	TOTRATE	x		
		Performing Arts Attendance	PERARTS (ALTERNATE)	x